Timeless ⋯ ⋯ ⋯

Copyright and Disclaimers

Timeless SEO secrets
3rd Edition

Copyright © 2021 Port Bell, Inc.

For permission requests, write to the publisher, addressed "Attention: Permissions Coordinator," at the address below.

Port Bell, Inc.
2602 S 38th St #361
Tacoma WA 98409
www.portbell.com

3rd Edition March 2021
First Edition: May 2019

TIMELESS SEO SECRETS
THE SEARCH ENGINES
DON'T WANT YOU TO
KNOW ABOUT
3rd Edition

www.seosecrets.info

By Dr. Ty Belknap

Timeless SEO Secrets

Table of Contents

Introduction

I am so excited to share the newly updated and expanded version of Timeless Search Engine Optimization Secrets with you. I have very specific SEO secrets to give to you, some of which the search engines don't want you to know about.

There is also a section on SEO Myths and a brand new section on *Timely* SEO Secrets. Over 40 pages have been added to this edition, and I hope it proves fruitful for you!

A quick note, none of what I have here is black-hat. I only do white-hat optimization (black-hat optimization isn't worth it in the long run).

And, just to be sure we are on the same page, there are no guarantees. This book is for informational and entertainment purposes only.

A bit about me: I'm just a techie nerd that's been sitting behind the keyboard, pounding away for the last 30 odd years working on websites and doing search engine optimization. I've been running a web design and search engine optimization company for over 15 years but decided to make a small change. I finished my Doctorate in coaching in 2017 and am now adding on business coaching and consulting.

A lot of what I do is techy-nerdy coding stuff that you wouldn't want to touch, and I use a lot of that to get to the top of the search engines. However, there are some things that you can do without knowing any coding at all, and they are included in this book. Yes, there is a tiny bit of coding involved, but I will walk you through it; the small bit of coding that will be in this book is easy to learn.

By doing all the secrets in this book, you will be far more likely to get to the top of the search engines without having to shell out hundreds or thousands of dollars on paid advertising campaigns. These secrets will be put directly onto your web site or will be things you do to help your web site. And, these secrets can *keep you on page one* for quite some time.

I designed my first website back in 1995. It was a web site on troubleshooting Microsoft Windows 95 (yes, I've been a techie nerd for quite a long time). And the website was really cool for those days (it's totally lame today but it was really cool for those days).

If you were around back then, you might remember that HTML 3 just came out. To give you an idea, we're now on HTML 5+. The powers that be have decided to stop creating "versions" of HTML and are now just constantly improving it.

Anyway, back in 1995, HTML 3 meant that you could actually put an image onto a web page without spending a half an hour to an hour trying to figure out the code correctly, crossing your fingers and toes, praying to God that it might work and 70% of the time still not working. That's what it took before HTML 3. It was a huge change. With this new version, it only took 3-5 minutes to get an image inserted onto a page correctly. Yes, *one* image.

That was back when the big search engines were Infoseek and Magellan. Maybe you haven't heard of those. Neither of them are around today. But there was one start up. A real pioneer in the industry at the time, but very small back then, called Yahoo®. Obviously, Yahoo has gotten a lot bigger while the other ones have gone away. One of the things that Yahoo pioneered was doing advertising on their search engine. The other search engines didn't do that very much back then because, frankly, not that many people really used search engines.

But Yahoo also advertised that they could help people get their web sites to the top of their search engine without placing ads and they charged a lot of money to do that. I thought if Yahoo can do that themselves, there must be some algorithm or something that would allow anybody to do it if they can figure it out.

And that's when I started stumbling upon some of these SEO secrets. A couple years later Google® came out, based off the DMOZ platform (which is probably way too techy nerdy for you. It's fine, you don't need to know about that). But Google got big quick, took what Yahoo did and exploded it a million times over and I started learning a lot more about how to work with the Google search engine.

But the funny thing is, many of the things that I was doing to get to the top on Yahoo also worked on Google. And some of those things I still do today, 20 years later.

In 2002, I started a web design and search engine optimization company. I have worked with over 200 businesses, getting over 3,000 keywords onto page one for the respective web sites.

I try not to work with competitors of clients, so I don't do many copies in an industry. Out of the 200 businesses I've worked with, I have probably done keywords for 190-195 different industries.

These secrets work no matter what industry you're in. You don't have to worry about someone trying to teach you something that's specific to one industry and hoping that it will work in yours.

I'm going to give you timeless SEO secrets, but here's the trick to it: **You don't have to be a techie nerd to do them.** Kind of cool, huh?

You don't need to know any coding to implement these tricks at all. As I said, there is a very minimum amount of coding, but I have already created videos or screen shots on how to do any of the coding in this book.

TIMELESS SEO SECRETS

Welcome to Timeless Search Engine Optimization
Secrets that search engines don't want you to know
about. You, reader, get to benefit from this book in a
way nobody else has yet. I originally created
Timeless SEO Secrets as a webinar and charged
$125 per person. Each time I performed the webinar
I learned a bit more, got great feedback from the
participants, and changed it just a bit.

Now you get to benefit from that experience. Yes,
this book is a lot less than $125, but it consists of
more than the secrets I taught in the webinar and all
of the steps to implement each one.

But a reminder; this is actual work. You're going to
be strengthening those finger muscles pounding
away at that keyboard. It's not magic, but there is a
formula to it because, if you do the right things in
the right way, it'll work. But you can't just sit
around, do nothing, and think that money is going
to be flowing in.

If you work hard at it, you should be able to raise your search engine rankings a great deal; whether you have a brick-and-mortar store, a small or medium-sized business, are promoting your entrepreneurial talents, or even if you are promoting a blog.

For those of you in the industry, these tips will help you shine whether you work as an internet marketing specialist for a company or own your own business.

There is only one thing you need for these secrets to work: A web site. Be sure to visit Kool Web Hosting (shop.koolwebhosting.com) if you do not currently have a web site. Just look for "Websites" and click on "Website Builder."

Black-Hat Vs. White-Hat

I'm looking for people with a strong work ethic to read this, and there's a reason for that; there are some techniques in here that could be used in underhanded ways. In the techy industry, this is called "black-hat."

If you're not familiar with the term, "black hat" actually comes from really old western movies around the 50's and 60's (yes, that old). In those days, the hero of the movie always wore a white hat and the villain always wore a black hat (that is, until Clint Eastwood came on the scene as the man with no name in "*The Good, the Bad, and the Ugly.*") That's where the term black-hat and white hat comes from. Everything that I'm going to show you will be the white hat way to do it, the good-guy way.

There are bad guy (black-hat) ways to do it as well and I will give you an overview of those, but only so you can avoid them. These black-hat techniques may benefit you for a short period of time, but eventually the big G and the big B will see what you're doing (those are the big search engines, of course they'll see what you're doing) and they will blacklist you eventually. If you get blacklisted, you will have to pay a specialist up to ten thousand dollars, maybe more, to get Google to let you back on their search engine. Once they see you are using black-hat techniques and ban your web site, you could pay a lot of money and it would take a lot of time. On top of that, you've only got about a seventy percent chance of success to actually get back into the search engines at all.

I'll give you an example: A client came to me saying they had been blacklisted because they hired an SEO person who did things the wrong way. They did black-hat techniques that got the company's main web site blacklisted from Google. It cost just over ten thousand dollars and took over a year of work; but I was finally able to get them back into the search engines.

And, by "back in the search engines," that means when somebody types their business name they could be found. Not their keywords, not what they did, just their business name. It took even longer to get them back to the top of the search engines for keywords.

By the By
If you type your business name (assuming it's not Acme, Johnson, or some other generic name) and you are not on page one of the search engines, there is something wrong with your web site.

And, along the same lines, if you hire an SEO expert that brags about his/her great skills by typing your business name and having it come up on page one, I would seriously think of freeing up that person's future. Getting the businesses name on page one is simple for an expert to do. It's like a high-end auto mechanic asking you to praise him for the amazing oil change he just did.

Remember, if your business is a competitive keyword, like ABC Plumbing, you may not be on page one. But you should at least be on page two without any SEO work being done. Look into hiring an SEO consultant if you are not, there could be a problem with your web site.

Not Get Rich Quick

For people in the industry, just to get this out in the open: This is not a get rich quick, homeless guy to gazillionaire story; even though when I was a teenager I actually did live on the streets for a while; and doing search engine optimization has changed my life financially a great deal.

However, you could make some serious money at it if you decide to do this as a business. I don't see why you wouldn't be able to make an extra fifty thousand a year once you get good at it.

And if you are a business owner who is trying to get your web site to the top of the search engines, I don't see any reason why you couldn't get a whole bunch more clients just from doing this (our clients average 20%-40% greater returns when we provide internet marketing services to them) because that's how people find you these days.

The best way to advertise is on the search engines, and this is going to get you there; **but there are no guarantees**.

Overview

So, are you ready to get to work? Are you ready to get those finger muscles all built up and strong? Together we will be in the Finger-muscle national championships and will rival Arnold Schwarzenegger.

These are search engine optimization super secrets that anyone can do. There is little coding; it's natural SEO, but some of it is techy-nerdy (sorry, can't help that).

If you're not familiar with natural SEO, it is when you go to Google and, at the top of the search engine you see the paid advertising; that's paid SEO (or SEM, Search Engine Marketing), you're paying to get to the top of the search engines. With natural SEO, you're not paying to have an ad at the top of the search engines; you're doing the work on your web page so it goes up in keyword rankings all on its own. You don't have to pay someone to stay at the top, you can do it yourself.

As part of this book, we're going to give you bonus free videos and another bonus: SEO myths debunked. You can watch any of the videos by going to www.webishops.com and clicking on SEO Secrets

In this book, *Timeless SEO Secrets* we, we will look at:

Blogs: Why blogs are more important than ever as well as how to install and update a blog for best SEO results.

Domain names: Chances are, you already have a domain name, but that's okay. I'm going to show you some domain name marketing tricks, even if your web site has been up and running for years.

Marketing web sites: How is your site marketed now? There is a way you can double or triple your web site marketing results, and it won't cost as much as you think.

Hosting: What does hosting have to do with SEO? A lot more than you realize.

Keywords: Keywords are dead, or are they?

Meta Tags: Okay, we are getting a little techy nerd here. But I've got you covered. I will help you get through this.

King of the Jungle: The domain name is the first thing search engines look at when indexing your site, but it is not the most important. Do you know what that is? Of course you do.

Never get Google slapped again: Don't know what that is? Don't worry, the full explanation is just a few pages away.

Freelancers: How to utilize the power of the internet's global reach to get you to the top of the search engines.

And more. But, on top of these SEO secrets, I'm also going to talk about several SEO myths that you want to watch out for.

Then I'm going to give you some *Timely* SEO techniques that are useful for now.

What Is Search Engine Marketing?

I'm glad you asked! Search engine marketing is one of those terms that seems as difficult to understand as the stock market, but it doesn't have to be that way.

Think of this; you decide to get your loved one something special for the holiday (pick whichever holiday you want) so you type "something special for my (wife, husband, significant other, dog, cat, whoever it may be)" in your favorite search engine.

What you just typed are keywords and companies selling products are trying to get to the top of the page when the results come up for those keywords.

To be more specific, let's say a water pipe breaks in your house, so you type in "water pipe repair New York (or whatever town you are in)." Or maybe you type "plumber New York," or "New York plumber."

Even though there are several words, that is still called a keyword.

Search engine marketing is the technique to get one web site above the rest when someone types in those keywords.

SEO Secret: Blogs

Blogs are ugly. They have hardly any cool moving images, and they are (if done right) full of ugly text that nobody cares about.

Search engines love blogs. They have all that awesome text that search engines can eat up and index. But, to get the most bang for your buck out of a blog, it should be done a certain way and in a certain place. Ready to dive into Timeless SEO Secrets?

The web site portbell.com has been up and running since 2015, and it does decent in the search engines. It ranks in the top 10 for about 50 different keywords. I added the blog in 2016. That is important, a year after the web site was developed, I added a blog. I post about 4 articles a month on the blog (transcripts for the Podcast, *Easy SEO and More*. Look for it in your favorite podcast player). In just over a year, the blog has had more keywords on page one than the web site.

Now, the blog would not do as well without the web site. The web site is the anchor for all your internet marketing. But, as you see, adding a blog makes a huge difference.

WordPress is the number one blog software platform (or app) right now. More people use WordPress to publish their blogs than any other software, and there is a reason for that: It's free and "relatively" easy to use (relatively meaning easier than most other web design platforms).

But here's the secret: most people go to wordpress.org to set up their blog. It's very easy to set it up with them; they do it all through their web site. But that helps WordPress a lot more than it helps you. Think of it, people are not going to your web site; they are going to the WordPress web site to read your blog.

The best way for you to publish your blog, especially if you are in the business of selling products or services, is to do the blog right on your own web site. It makes sense when you think of it, and most of these secrets are about simple things that make sense. It's just one of those little things that most people don't do because they don't know why they should.

So, don't publish your blog on wordpress.org. And don't get me wrong, publishing it there might help a bit. But it's going to help a hundred-fold if it's on your website. Think about it, you have taken the time and trouble to set up a blog. You are consistently posting articles (I will talk more about that in a bit); and you put all those keywords, tags and categories on that blog so people will find it. But, since the blog is not on your web site, it isn't helping your web site rank higher in the search engines; it's helping WordPress.org. Yes, some people are probably going to your site, but the goal is to get you higher in the search engines.

Want to set up a WordPress blog or web site the easy way? Go to **www.koolwebhosting.com and click on WordPress Hosting**. The system sets everything up for you, creates backups, and keeps you up to date. But what's most important is that you can connect this blog to your web site, even if your web site is hosted somewhere else!

SEO Secret: Blog Categories

Look at a blog. If you can't think of one offhand, go to www.forusintroverts.com, and look at the right side (or take a look at this screen shot):

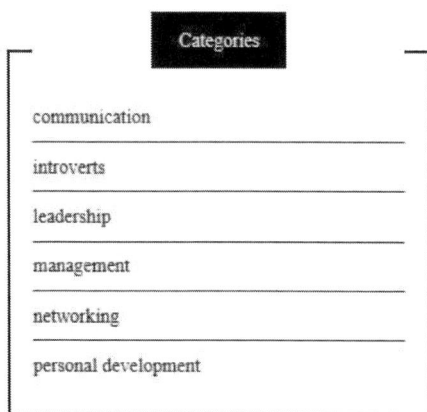

Categories
communication
introverts
leadership
management
networking
personal development

These are categories that I created so people can find relevant articles easier. But, they are also there so search engines can index them. One secret about blogs is that every post is treated as a separate page in the search engines. And each post can have its own categories, keywords, and Meta tags (both Keywords and Meta Tags have their own Secrets areas below).

It is important to put those into every post you create. Remember, your posts are for people, but people won't find them if the search engines can't index them.

And as I said, search engines LOVE blogs. The entire function of a search engine is to properly index text; and blogs are huge blocks of text. So, try to do a blog post once a month. Don't have time? Don't think you will remember to do one every month? Well, there's a secret to that also. Keep reading.

I am not going to get into the details of how to install WordPress, there are many videos already out on that. Once it is installed, you should add some plugins (again, there are many videos on that). I would recommend an anti-virus (WordPress is easier to hack than standard HTML pages), a program like Cerber (it takes care of unwanted spam comments), and an SEO plugin like Yoast.

Next, start writing posts. Here is how I recommend doing your blog posts:

1. Fill in the title with at least one good keyword (see the section on keywords if you need help).
   If you are a landscaper in Los Angeles and you are writing a post on how to dig a ditch the easy way, try a title like "Los Angeles

Landscape Ditch Digging...," or something similar. Yes, it is a bit long, but remember that we are doing this as much for the search engines as we are for people.

2. Write the content. I would recommend 2,000 words. No, that's not a typo. There was a study in 2018 that found the average web page that was at the top of Google was over 2,000 words. No, every blog post does not have to be that large, but they should be if you want to be in the top of the search engines.

   Try to incorporate your business name, what you do and where you are at least once a quarter (one in every three posts). And don't be afraid to put a contact us link at the bottom of the post.

3. Add an image. Take a picture of a ditch you recently dug, or a ditch someone else dug that wasn't done well. Try to make the picture fit the post, but add a picture. Statistics show that blog posts with images are 30% more likely to be read.

4. Add categories. For the Los Angeles landscaping company, categories could be something like: Los Angeles (your city is not a bad idea), landscaping (what you do is a

must), ditch digging (add identifying categories if you plan on doing multiple posts about a subject), etc. Don't skip this; most people don't use the categories option. That means that you will be ahead of them.

5. Add keywords. Keywords are an option right below categories (this is another one most people don't take advantage of; run by your competition in the search engines and win the race). This is where you can get creative, but don't go overboard. Don't put keywords that do not relate to the article. People care if a keyword is not relevant; and you want people reading your articles.

Blog writing secret: I do a blog post pretty much weekly. Most are small posts, but some are large (2,000+ words). I have a tough time sitting down every day or two and writing a blog post; I'm much better at writing marathons than sprints. So, I sit down and write 4-8 blog posts at one time. Then I *schedule* (you will see the option next to "post now.") the posts to go out over the next month or two. That way, I only have to write posts once every month or so. But I do keep track of post ideas throughout the month, so it is easier for me to come up with ideas.

If you have a job that makes it difficult to keep pen and paper, I'll bet you have your miniputer nearby (come on, they stopped being cell phones, or even smart phones, a long time ago). And they have these cool little things called apps. Install a notepad, use your voice recorder for notes. Or, even easier, write yourself an email and create an email category called blog ideas. I'm sure you can come up with *some* way to keep track of your ideas. Do what works for you.

And, just in case you need it, here is your excuse to get a motorcycle. I come up with many of my blog post ideas while cruising back-country roads. I like having my notepad app because it doesn't matter if I have a cell signal, it will still store my ideas. I just stop on the side of the road, leave myself a quick note, and I'm off enjoying the day again.

If all else fails, here is another **Blog Secret**: Write about magazine articles you read. Do you tweet? Good. No, I mean having Twitter is good. But not why you think. And if you don't have Twitter, you may want to get it because you can follow magazines on Twitter. They post every article they write and when you follow them, you get to see their latest articles. Read one while you are waiting for heart transplant surgery. Well, maybe while you are waiting at the doctor's office for your checkup. We spend a lot of time waiting, take advantage of it.

Then, when you read an article you think would work on your blog, post a link to it and make your blog about your take on the article. What did you find interesting? What would you do different from what the article says? Write about it.

Still think you don't have the time or ability to write blogs? Don't worry; I have you covered in the SEO secret on Freelancers.

SEO Secret: Social Media

I also have a SEO Myth about Social Media, be sure to read it. But here is an SEO secret that came out in 2020: **Social Media Is Showing Up In Search Engines**.

This is huge and not a lot of press has been coming out about it. It won't take long, but for now this seems to be a secret. Facebook posts are starting to show up in search engines. So far, I've only seen posts from business pages, but I've seen them on both Google and Bing.

Also, I was doing a Bing search recently on SEO, and my video on "What Is SEO?" popped up. Now, that is not strange on the surface, but *it wasn't the YouTube video that popped up, it was the video I uploaded to Facebook!*

So, start uploading those videos to Facebook and be sure to add in good descriptions so they will show up on Bing (is Bing trying to battle against Google and YouTube?).

LinkedIn articles are also starting to pop up on search engines. As far as I can tell, posts are not showing up yet, but articles are. If you aren't familiar with LinkedIn, it's a business to business (B2B) social media platform. LinkedIn allows people to create posts for everyday updates and articles for things that have a more permanent nature.

So, start creating articles for LinkedIn and videos for Facebook. These won't help your web site, but they will help your business/organization/team/etc.

But keep keywords in mind. You're now competing against the millions of other social media-ites for the top positions in the search engines.

SEO Secret: Hi, My {Domain} Name Is···

Domain names are the very first thing that search engines look at when they index your site; and what they find will help them determine what keywords they are going to promote. Now, domain names are not the most important thing search engines look for, but they are the first thing; so, your domain names are important. That is why I recommend getting **marketing domain names.**

My SEO company is Port Bell (www.portbell.com). But that domain name doesn't say a lot about what I do, which is common among company names. They create generic names (Amazon.com, Apple) or names about the founders (Hewlett-Packard, Proctor & Gamble) so they can offer multiple products without worrying about changing the company name.

Rather than make up some fake company name, let's look at what I did wrong with my web site, www.portbell.com. It is an SEO web site, but the domain name doesn't say anything about SEO.

That doesn't mean I couldn't get the web site to page one, it is for many keywords. But, like I said, the domain name is the first thing search engines look at so it's good to have a name that explains what you do.

So, I created a series of marketing domain names. One of the names I created was puyallupseo.com (Puyallup is a small city in Washington State). That name is nice and long. Some may say that at least it is a .com, but that does not matter at all.

A marketing domain name is created for one reason only: Search engines. People will not be typing the name in; it is only for search engines to find, so make it (almost) as long as you want. I wrote almost because domain names are limited to 256 characters (still gives you a lot to play with).

And, since marketing domain names are not really for people, you can get away with long names. Sometimes a longer name is even better if you can put a couple extra good keywords in it. You can even use .info, .us,. whatever you want because search engines don't care what the extension is.

A San Antonio plumber that specializes in digging ditches could have a marketing domain name of plumbingsanantanioditchdigger.com. That is a long name, and no good for a company's main site. But it's perfect for a marketing site.

Are you starting to think of creative names? Find out if they are available. Go to **www.koolwebhosting.com** and see what you can find. Search for different types of marketing names and that site will tell you which extensions are available.

Maybe you want your company name in every marketing domain name. You could put mycompanyplumbingsanantonio.com.

Here is the Super-Secret of Marketing Domain Names:

You might be tempted to forward your marketing name to your main site, and that is not necessarily a bad thing. But some people used to use what's called "masking" back in the old days.

Here is a quick definition of masking if you do not know what it is: Masking allows more than one domain name to point to the same web site without letting people (or search engines) know that they are both pointing to the same web site.

For SEO, This is Black-Hat. Search engines know about this now. They will see two different domain names, but they will see an exact copy of the web site so they will think you are trying to spam the search engines. This is not a good thing. ****Do not use masking unless you know what you are doing.****

Just to be clear, forwarding is good and fine. Just don't use the masking option (it's a bit complicated to use anyway).

But, forwarding can be a bit techy-nerdy, so I have created a video on how to quickly and easily forward a domain name. Just go to **www.webishops.com** and click on SEO secrets. You will see the video there.

Instead of using masking, the best thing to do is to create a separate marketing web site.

SEO Secret: Marketing Web Sites

Marketing domain names can give you a great advantage over the competition, but domain names themselves don't do much. So, in order to really take advantage of that keyword marketing name, the best thing to do is create a marketing web site.

A marketing web site is designed exactly for what the name implies; marketing. If you sell shoes, and the latest summer shoe is coming out soon, and you just know that everyone will be looking for them, create a marketing web site just about how you are the person to go to for those shoes. If you are a real estate agent that wants to sell a high-end house, you could create a marketing web site just for that house.

Here is the first secret to a marketing web site: It can be as little as one page. Web designers will tell you that a one-page web site will never get to the top of the search engines. That is not true in the least bit. It may be difficult if it is a competitive keyword; but it's not impossible. However, I do recommend a marketing web site of at least three good pages of content. Don't put a contact page on there, or an about page, or any of that; you don't need it.

"WHAT? No contact page? How are people going to contact me?" I hear you thinking already. You won't need a contact page. Remember, this marketing site is designed to get you to the top of the search engines for a particular keyword. But you don't want the people to stay there.

All the navigation of your web site (except the three marketing pages, of course) should link right back to your *main company site!* Get the people off the marketing site and onto your real web site as quickly as possible. They should have everything they need to contact you on there.

However, that being said, you should have your phone number on each page if you want people to call you, and a link to your contact page should be on each marketing page. That way they don't need to go searching to find you.

The secret of Marketing Web Sites: Make your marketing web site look exactly like your main web site, including the navigation.

Copy the *look* of your main web site to the marketing site, but you want to change the *words.* It is vitally important to make sure the content is 100% original or the search engines will think the site is spam and all your effort will be for nothing.

Check out the section on "King of the Jungle" for more information on content. And don't worry; this can be done affordably if you don't want to write the content yourself (hint: See the SEO Secret on Freelancers).

Yes, creating a marketing web site can cost some money. A freelancer may ask for $400-$600 to create a marketing web site for you; and that might be steep if your goal is to market a new line of shoes for three months, so here is...

Another Marketing Web Site Secret: Website Builder. Go to **www.koolwebhosting.com** and click on "Build a website." The Website Builder page should pop up. This is a very easy way to get a web site up and running, whether you need it for three months or 30 years. Choose the Business option. Yes, it is a bit more money, but it is responsive (meaning it will automatically shrink to smart phone size and grow to desktop size).

Example: You have a product that is a bit older. But you have a lot of them and want to get rid of them quickly, so you decide to sell it at a great price. You want to highlight that product so you can sell it fast and decide to create a quick marketing web site.

You pick a good domain name that has keywords (thisproductforsaleinmycity.extension) and want a web site, but $400 is more than you want to spend to hire a freelancer. You know that, if you can get the word out, you can sell that entire product in 3 months.

Use Website Builder. Let's break it down:

You purchase the keyword marketing domain name for under $30 (it should be a lot less because, remember, the extension doesn't matter. Pick the least expensive extension you can find)

You purchase Business Website Builder for three months (you can always add months later if all the items don't sell) for under $30.

It will cost you well under $60 for the domain name and web site.

It may take a few hours to get the web site set up completely; but when you are done, you are out under $60 and five hours of work (and remember, there is 24/7 support if you get stuck).

Once you have your marketing domain name and web site ready, you need a place to host it (this does not apply to Website Builder, hosting is included!).

SEO Secret: Hosting

What does web site hosting have to do with SEO Secrets?

I'm glad you asked!

(Remember, you can go to www.koolwebhosting.com and click on Web Hosting at any time to look at hosting options)

Hosting is the address where your web site resides, much like an address for a house or a business. Hosting addresses are IP addresses (something like 192.168.0.0). And there are several types of hosting:

Shared hosting is like a mail drop box. With drop boxes everything goes to the same address, but sometimes with a different suite number. However, with the popularity of day offices, it is now possible for several businesses to have the exact same street address.

And that is what shared hosting is. Everything goes to the same address, and then a virtual secretary in the hosting software determines which web site to show you based on what you typed.

There was a time when search engines would not index a group of web sites if they had the same address, because they thought the sub-sites were spam. That is not true anymore; you can have several sites in a shared hosting environment, and they will be indexed by the major search engines.

However, shared hosting still does not do as well in the search engines as dedicated hosting.

Dedicated web hosting is like a house address. Only one house can be at a certain address. And with dedicated hosting, only one web site is at a certain IP address.

The advantage of dedicated hosting is that search engines like it better. I would say that this may change, but it's been this way for over 20 years now (for no good reason I can think of).

The advantage of shared hosting is that it is more cost effective; you can have many web sites on one hosting account. For instance, as of 2020, you can get a dedicated web hosting account at www.KoolWebHosting.com for under $6 per month. However, you can get a shared hosting account that can hold up to 100 web sites for under $10 per month.

And here is the Timeless SEO Secrets Super-Secret: Just for reading this book, you can get 15% off any order over $50 at www.koolwebhosting.com. All you need to remember is to use the code "seosecrets" (without the quotes) at checkout. You can use this code any time you want.

Don't go with the free hosting accounts. Actually, I won't say don't; do your homework when you go to the free hosting accounts. Some of them may be okay, but some of them are a little bit trickier.

I worked with a client who had a free hosting account. We designed a new web site for her, so we needed to move her domain name (which was part of the free hosting account). However, the company wanted to charge several hundred dollars to release the domain name since they had allowed her to use it for free.

They had all the power; they actually owned her domain name according to their policy. So, her choice was either to give up her domain name and start from scratch or pay a couple hundred dollars to keep the domain name that people already knew.

Pay careful attention to the costs to move your web site before getting that "free" domain name and hosting.

SEO Secret: Keywords

Did you come to this secret first? It's okay; it's not your fault. Like Pavlov's dog, you have been conditioned to drool at the mouth and open your wallet whenever we SEO people speak the name: "Keyword" (and you can send the check to…).

However, unlike many other things, keywords really are gold and they are a part of the King of the Internet Jungle.

Keywords are very important. This is where a lot of people goof up because they will do a website all about their company, writing things like "Hey, we're XYZcompany. We've been in business for 125,000, years. We are amazing! Look at this picture of our owner! Our owner is a great guy. Look at those pictures of his family; he's got a great family. They do volunteer work in the community up to 185 hours a day…"

I mean, they go on and on and on. They haven't said a single thing about what they sell or anything about why people should buy from them, but there sure is a lot to read and look at.

What do you sell? Where do you sell it? Who will benefit from it? Why are you different from your competition? The answers to each of those questions are keywords.

Are you a life coach in Portland, Oregon? Start with local keywords before you look at national keywords. The front page of your web site, the signature on your emails, and your internet marketing should start with marketing your service in your hometown.

Maybe you live in Portland, Maine, a beautiful city of 60,000 people. Your thought may be that you want to market to a lot more than 60,000 people, but how many of those people would you need to work with before a boatload of money starts coming in? Even if you sell a product that can be purchased nation-wide, start locally. Most big businesses got big by starting locally.

And I'm only harping on that because your locality is part of your keyword structure.

So, look at your locality and see what you can do with it. For instance, I live in Washington (state). Washington is one of the most common names (for cities, counties, schools, etc.) in the country. I would not do keywords with the name Washington. However, there is not another city called Seattle anywhere in the world. Plus, anyone who knows the name Seattle knows it's in Washington.

Fairview (the #1 most common city name in the world) would be difficult to use it as a keyword. But you could use Fairview, (yourstateabbreviation). Or, what is around Fairview that is unique?

Don't get too creative with your keywords. Remember, you want keywords that people look for. I worked with a church that used a cryptic # keyword for their posts. They liked it because nobody else used it. However, nobody else looked for it either. I told them a good keyword would be #tacomachurch (they were in Tacoma, WA).

Can you get too pinpointed with your keywords? Another client had a restaurant on Pacific Avenue (a quite common street name on the West coast). And he specifically wanted to get to the top of the search engines when anyone was looking for a restaurant on Pacific Avenue. He did keywords for the city and his specialty meals also, but he was adamant about being at the top for Pacific Avenue. I had to ask him how many people he thought he would attract with that keyword; and I loved his answer. He said if ten people a day found him from that keyword and came into his restaurant, that's ten more people every day and it's worth it. So, we did it and it worked fantastic.

You can use your business name as a keyword, but it will only be good for people that already know you. Remember, nobody cares who you are until they know about what you can do for them. The benefits of your product or service are great keywords.

Be sure to use of every kind of iteration you can think of for your keywords. Plumbing, plumber, plumbed(?). Ditches, draining, pipes, house pipes, water pipes, etc.

Think if every word that you can and combine them with your city name or locality.

Here is the Keywords Super-Secret: When you use your keywords with your local area, you are automatically creating national keywords at the same time. If you have the keyword "gift baskets Wichita," you are coding for that keyword as well as "gift baskets." Work on locality keywords, the national keywords will work themselves out.

SEO Secret: Meta Tags Are {not} Dead

Yes, META tags are still useful in your website. They don't have the weight they once had, but think of it this way: You don't have to wash your car, but it looks better and will be less likely to rust if you do

Meta tags are a subject that I love because Google, not too long ago, came out with a whitepaper saying that Meta tags are dead (oh, by the way, a whitepaper just is an official document by a company). The whitepaper detailed how Google doesn't do Meta tags anymore; they don't index them. There were four or five paragraphs on why they don't even look at Meta tags anymore. And, at the bottom of the whitepaper there was a link that said, "click here to effectively use Meta tags on your website." To say that Meta tags are dead is actually a myth (it's not part of my myths because I felt it was more important to have it in this section).

There are several different types of Meta tags, and you should have Meta tags on every web page on your site. The three most common Meta tags are the Title, Description, and Keywords (yes, you can put your keywords in this Meta tag, but don't use a word more than once and don't use commas or semi-colons to separate them). Here is a sample of what Meta tags look like in the code section of a web page:

<TITLE>SEO Tacoma Seattle Internet Marketing - Intra-Designs. Keep your search engine optimization local by Port Bell</TITLE>
The TITLE tag is the main text you see when searching for something. The TITLE tag can be up to 120 characters. The max used to be 60 (and you will still see anything over 60 as "..." in the search engines at times), but times are changing and search engines are starting to show up to 120 characters. If they will show that many, use the space.

<meta name="Description" content="Affordable Seattle SEO Tacoma Search Engine Optimization Olympia and internet marketing for Western Washington at Intra-Designs, Inc. Tacoma SEO. Get more customers by getting to the top of the search engines. Contact us today. 253-44-55-777." />

"They," meaning people that work for search engine companies, used to say that a description tag should not be more than 120 characters. Now 240 is the average.

<meta name="Keywords" content="SEO tacoma search engine optimization seattle web design puyallup internet marketing coach website ecommerce consulting sumner bonney lake lakewood puget sound pierce county washington" />

"They" (see above) keep telling me that this is the section that is dead, but it keeps working in the search engines so how is it dead? There is no limit to what you put in here, but it does not carry as much weight as it used to.

I am not going into detail on how to put them on your web site, there are many videos and instructions already out on how to do that. But you can copy these tags and put them into your web site. Be sure to only change what is in between the quotes in the content=" " sections of the keywords and description tags.

We can't always take at face value what "they" say because, several years after I was told "META tags are dead," "they" came out with about 15 new META tags. You can now have specific META tags for when your web page is linked from a social media page, but that is more than I'm going to get into here.

You can put these three Meta tags onto your WordPress blog posts as well if you have a plugin like Yoast SEO. However, instead of a Keywords Meta tag, turn your keywords into categories and tags (you will see both of these as options on the right side).

However, follow these steps to use them with each blog post.

Your screen should look similar to this when creating a blog post:

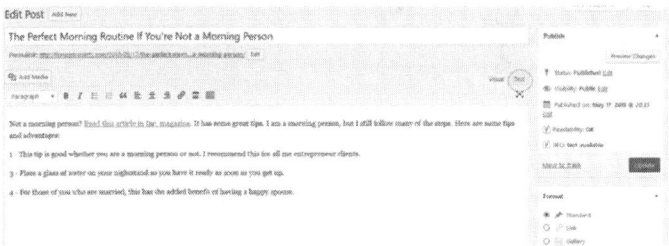

See where Text is circled? Click that, and then add the three Meta tags at the top, before anything else.

Looking at these three Meta tags, you can see how using keywords will help you get to the top of the search engines. But pay special attention to what you put where.

A Meta Tags Super-Secret: Keep title tags under 120 characters. Keep your description under 240 characters.

SEO Secret: King of the Internet Jungle

Are you noticing a theme yet? So far, two of the SEO secrets (Blogs and Marketing Web Sites) have talked about the importance of content. Well guess what? The domain name may be the first thing search engines look at, but it's not the most important.

The King of the Internet Jungle is content. If you don't go away with anything else, this alone can get you above 70% of the other web sites in search engines. Search engines love content; and they love content that changes even more, which is why they love blogs. If you're writing a blog every week, search engines will be indexing that blog constantly. It is estimated that search engines will look at a web site 10-50 time more than it changes. If you change your web site once week, search engines will look at it up to 50 times a week. If you change your web site once a day, the search engines may look at it up to 50 times a day for changes.

So, think of a web site that you have gone to that looks cool; maybe it has a video running at the top and some amazing imagery right below. That web site may be cool, but the search engines will ignore the video and the images. All search engines care about is quality content. That is why boring blogs help so much.

I recommend changing the content on your web site every week. And no, I don't mean the whole site or even a whole page. Add "title" text to an image, put "alt" text on a hyperlink (these are pretty easy to do in WordPress (and, if I just started talking techy-nerd-speak, do a search. There are tons of pages that explain these). Change the wording on a paragraph, add a sentence or two, or remove a sentence. Search engines like web sites that change, but they don't care if those are small changes.

Change a different page each time; you don't have to change all the pages at once. In fact, only change all the pages at the same time when you must. And, when you make changes, try to do it on different days each week and different times.

I found this out by accident. I scheduled Thursday morning to do SEO work on a certain web site. Every Thursday morning, I would make changes. Well, one Thursday I was sick, and I didn't get it changed until Sunday that week.

The web site shot up in rankings within two weeks. Surprised, I did some research. It turns out that the search engines would come and index my site 2-3 times every day to see if there was a change. However, since I missed Thursday, and did it on Sunday afternoon instead, the next week the search engines were coming to my site up to 10 times a day to see if anything had changed.

The more search engines look at your site, the more they're going to be indexing your site, the more likely you are to get to the top of the search engines.

So, as an experiment, I decided to start making smaller changes twice a week on different days and at different times. The search engines started coming to that web site up to 50 times a day.

The Content Super-Secret: There is nothing else you can do to get search engines to come to your web site looking for changes 50 times a day. That is how important changing content is.

Of course, by search engines, I'm talking about the spiders. But Internet spiders are nothing like real spiders; you don't have to be afraid of those. In fact, those are actually things that you should like. Internet spiders are just little automatic programs that go and check web sites to see if anything has changed (they are not much different than phone apps).

The more often you change your web site, the more often the spiders will be looking at it. Content is King. It is the number one way to get to the top of the search engines. If you do nothing else, and you change your content every day, you could still outrank other web sites.

SEO Secret: Slap Google

Now, don't worry if you've never heard of Google Slap before. If you haven't heard of it by now, it's probably something you won't need to worry about. And I've actually already written about Google Slap, but I didn't call it that.

A quick definition of Google Slap: Think of the scenario I talked about in the Secret on Marketing Web Sites. XYZ company has a main web site, and they decide to create three marketing sites. But they don't use original content; they just copy the content for certain products onto the marketing sites. But they really want to promote those products, so they send the same content out to other people who create web pages on their own sites (a common practice in some industries).

But the content is still exactly the same. Google calls that Spamming. It is Black-Hat (even if it is unintentional), and they will eventually blacklist every site they find with all that duplicate content. Blacklist, if you remember, means each of those web sites is completely gone from Google. Slap!

NOTE: Keep in mind, this only applies if a majority of the web site is exactly the same as another web site. Having one page exactly the same as a page on another web site will not blacklist you, but it will not help you one bit either. Google will most likely ignore it, knowing that the page on the other site is the original.

The Slap Google Super Secret: Create 100% original content for your marketing web sites. Now, some of you may think that this is no secret at all. In fact, you may say it's a "duh." But, especially if your marketing web site is a copy of your main site, did you change the names of the *images*? How about the Meta tags? Did you change the name of the pages? Not just the title, the actual page (HTML) name. Change the name of everything you can and you will never get Google slapped.

SEO Secret: Freelancers

This is the **Super-Secret** for all you people whose eyes are crossing about now. I know, there isn't much coding, but it does get techy-nerdy. So why do it yourself? Hire someone else to do it for you. Hire us, we can work with freelancers for you so you get what you want every time and don't lose money. Go to **www.portbell.com** and click on Contact Us.

Of course, you can do it yourself as well. Hire a freelancer in India, Pakistan, Hungary, or even in the good ole USA.

There are two great websites that I know of where you can hire freelancers. One is upwork.com. The other one is fiverr.com (with two R's).

Whichever one you choose, please do your homework. Pick a designer or coder, whatever you're going to be hiring, that has great reviews and has a lot of reviews. Make sure that they guarantee their work. That way, you can get your money back if you don't like what they did.

There are flaky people on these freelancer sites, and sometimes a reliable person just has issues out of their control. I was working with a young lady in Serbia a couple of years ago when the fighting broke out and she disappeared. I do hope she is okay. However, because of her guarantee, and the guarantees of the two web sites I mentioned, I was able to quickly get my money back and hire another person.

Here are my tips for working with a freelancer (whether on these sites or somewhere else):

- *Be specific about what you want. Tell the freelancer exactly what you want. Include screen shots, samples of other people's work, anything you can to show them just what you want. Detail everything you can, even the date you want it delivered by. Make sure this is in the initial paperwork before the project is started so you don't surprise them.*
  This is not always the best thing to do if you are asking for creative work. I once asked for a specific design; but I had a great designer. He came up with a better design and showed me both what I wanted and what he came up with. If it is a creative work, let them be a bit creative.

- *Get a guarantee. Have them guarantee their work. And make sure the guarantee is enforceable. Check to see what the site refund policies are.*

- *Never give the freelancer money until it is 100% done. Put this in the initial agreement. Freelancers will ask for anything they can up front, but you have no way of knowing if, or what, they will deliver.*

- *Use an escrow account (required on upwork.com and fiverr.com). This lets the freelancers know the money is there and waiting for them after they deliver what you wanted. If they don't deliver, you can use the money to hire another person or have it returned to you.*

There are differences between the two freelance sites:

Upwork.com is generally more expensive, but the freelancers are generally of a higher caliber. You are more likely to get exactly what you want, but it will cost more.

Fiverr.com is less expensive (many gigs can be purchased for as little as $5, hence the name), but the freelancers are generally not as good. But that is relative also. Fiverr has some excellent freelancers (the cover of this book was done by a Fiverr freelancer), and I have had as many issues at Upwork as I have at Fiverr, so there are no guarantees you will get what you want the first time. That is why it is so important to use an escrow account.

SEO Secret: Backlinks

Backlinks are links that point to a page (or blog post) on your web site and backlinks have a long and very involved history. There was a time when the more backlinks you had, the better.

But then the search engines cracked down because there were people posting stupid (yes, stupid) posts onto every blog they could find with irrelevant backlinks. So, spam backlinks started getting sites blacklisted in Google and Bing.

But then people complained, because they had no control over who created a back link to their site (some competitors used this to get ahead of others). So, the search engines had to change and start rating backlinks (sound like some of the drama shows out there?).

You won't get blacklisted for having a spammy backlink to your site now. It won't even matter if you have 100 spammy backlinks to your web site, but it won't help either. The search engines now rank backlinks using many factors; one of which is the likelihood that you actually know the person or company that is linking to you.

Popularity is another factor in the backlink game. The more popular the web site that is linking to you, the more important that link is to the search engines and the higher your domain score will be.

SEO Secret: Crosslinks

Which is why crosslinks are the way to go. The best backlinks are those where both web sites link to each other. But Google even ranks these links. Yes, bad crosslinks (if there is such a thing) are better than above-average backlinks, but there are ways to get even more mileage out of crosslinking.

1. **Share links with a local business.** Google knows the physical location of businesses and can tell when local businesses link with each other.

2. **Share links with competitors.** Sounds strange, right? But a plumber in Seattle, WA isn't really a competitor to a plumbing company in Miami, FL. Share links with competitors that are out of your target zone. Even if you run an online business, and the world is your customer, you can crosslink with your competitor. Chances are, you don't do exactly the same thing as all of your competitors. A good example: I do online marketing, but I concentrate on SEO. Amy Porterfield does online marketing (she even has a popular podcast on it), but she concentrates on email list building. These are

two very different niches in the online marketing arena.

3. **Share links with complementary businesses.** What business is complementary to yours? There are always complementary businesses. An artist could share links with the art supply store they get their items from or where they sell their artwork. An apartment complex could share links with construction companies, roofers, etc.

Make a list of 10 types of complementary businesses, then find a company in each category and ask them to share links with you.

SEO Secret: Mobile Friendly

This used to be a myth, but now that myth is a secret. Or at least part of it is. Part of it is still a myth.

What is mobile-friendly: Mobile-friendly means that a web site will change how it looks to fit on a mobile phone. You may have gone to a web site on your smart device before and had to scroll left-to-right to see everything. That web site was not mobile-friendly.

A mobile-friendly web site will usually have larger font sizes when on a phone (so you are not squinting to read the text), picture sizes will be smaller, and you won't have to scroll left-to-right to see everything.

The Myth: It is still a myth that your web site **must** be mobile-friendly to rank high in the search engines. Not every search engine requires this yet. In fact, as of early 2021, Bing, Yandex, and most other search engines do not require a web site to be mobile-friendly.

Even Google does not require a web site to be mobile-friendly. However, a mobile-friendly web site (even if it's worse than an old-school web site) will rank higher. So, it's worth it to have your web site mobile-friendly

The Secret: There are now two different types of mobile web site design (I know, we're getting a bit techy-nerdy, but it won't be too bad): Mobile-Friendly and Mobile-First.

Have you noticed that sometimes when you go to a web site, you will see three lines instead of a menu:

Clicking or tapping these three lines will bring up the menu. If you only see these three lines, even if you are on a regular or laptop computer, that is a **mobile-first** web site.

If you see a normal menu of items on a regular computer or laptop but see those three lines while on a tablet or phone, that is a **mobile-friendly** web site.

For marketing purposes, do not have a mobile-first web site. This has little to do with SEO, but it has everything to do with your customers. Some people will be confused with the three lines and think that your web site is only one page.

Timely SEO Techniques

Most of this book talks about timeless SEO secrets, and the advantage of those secrets is that they have been useful for many years.

In fact, Google® has done a major update in March of 2021 and, for the first time, they are warning us of the update. **But the update addresses nothing more than what's already in the pages of this book.**

Google wants us to create original, compelling content. Period. Do that and you will get higher in the search engines.

But there are some techniques that are also working "for now."

SEO Technique: Lead Magnets

Lead magnets, if you are not familiar with the term, are useful pieces if information that you give away for free in exchange for someone giving you their name and email address. A lead magnet can be almost any form of electronic information: PDF, video, audio, informational emails, etc.

Go to **www.tysfreebook.com** to see a sample of a lead magnet.

Lead magnets are a great way to get new customers, but they can also be used for SEO purposes. If you notice, the link above goes to a web page where you can download "50 Marketing Strategies for Entrepreneurs." But that page does not have a lot of text.

That page could be better utilized for SEO by having a lot more text (and yes, that is a page on my site, I'm not perfect either). But here's the difficult part: How do you put a lot of information on a lead magnet page without giving away everything in the lead magnet?

Lead magnet pages do not have to be 2,000 words. And, in fact, they probably shouldn't because your lead magnet probably isn't that many words. Use enough words to describe the lead magnet without giving everything away.

For my lead magnet, I could give away one or two of the 50 strategies without lowering the value of downloading the lead magnet (and I may have already done that by the time you read this).

So, do your best to use lead magnet pages in a way that will attract search engines, but there is one big disadvantage to lead magnet pages: They have (or should have) no links to any other pages.

Search engines like web pages that have links to other pages (internal links to other pages on your web site are just as important as links to other sites). This will make the page less valuable to search engines, so it probably won't get as high as your other pages. But that doesn't mean it won't get into the search engines.

SEO Technique:
Keywords in Page Names

This is an SEO technique because it's already starting to lose weight in the search engines, but it is still important for now.

Make your page names descriptive and relevant. Some pages on your website have to have a specific name, **have to,** and for business purposes you should standardize some of your page names because remember; our job as businesspeople is to get people to your website.

Even if you have a creative business, you do want to standardize some page names. For instance, your main page should always be index (or Home in WordPress). that is definitely kind of techy- nerdy. But, it has to be that or people may not see your main page when they come to your site and search engines may not index it properly.

History: Way back when, during the dark ages of the internet, when there was no such thing as a web site, all file names had to be eight characters or less. You could not have a name like "29-seo-tips-for-2021." You would get an error and the file would not be uploaded to an Arcnet (what the internet used to be called) server. Because of that, most filenames were very cryptic. You could have a file like tpsy101a.html.

But what would a name like that mean? Well, every area on the web server had a main file name called index.html. And, just in case you haven't figured it out yet, that file was an index of all the other files in that directory.

It would say things like:

tpsy101a.html = Topic: Psychology 101 assignment A.

The index page is a legacy file that had become the default main page of every web site for the last 50 years. But times are changing. "Home" can now also be used as the main page, as long as there is no "index" page. But, as of now, those are the only two words you can use to name the main page of your web site, or any directory in your web site.

Some other pages can be more creative. Instead of contact for your contact page, you could have contact-seattle-plumber. That makes the page more descriptive for both people and search engines. See the SEO Technique: Page Names for more help on all other pages in your web site.

I'm talking about the URL, or permalink, of the page, not the META title or main headline you see for the page.

SEO Technique: Page Names

Use a – (dash) instead of a _ (underscore) or a space in your page names (URLs).

Yeah, this is kind of techie nerdy. Sorry, but it is important: Always use a - between words in your page names. For instance, instead of SEO_company, SEOcompany or "SEO company," use SEO-company. Using a - makes it easy for search engines to separate the words. Using an underscore makes it look like all the words are running together.

And never use a space, the internet still sometimes interprets a space as %20 so SEO space company could be seen on the internet as SEO%20company, which means nothing to the search engines or to people.

History: Way back when, in the beginning of time (or 30 years ago for those who remember a time without computers), computers could not recognize a space in a filename. The computer would come back with an error any time a space was used, and the file would not be saved, so all that hard work would be gone. So, some smart person at Microsoft (because this was a DOS issue, not an Apple 2 issue) found a way to automatically fill in that "space" with what is called ASCII coding. Now, the ASCII code for a space (basically) is %20, which is why we still see that sometimes in page names on the internet. What is surprising is that this is very old coding and should not be seen anymore, but we still do sometimes.

SEO Technique: Canonical... WhoWhat?

Okay, this is very techy-nerdy, but it is also becoming quite important. I listen to a lot of technical podcasts about SEO, and this has come up on three of my podcasts within a 30-day period of time.

Historically, duplicate pages on the internet were very bad and could lead to a web site being banned from search engines. But this became difficult to manage with all the media outlets today. It is now possible to have one news story copied word-for-word on over 100 web sites.

So, a META tag called "Canonical pages" was created. It is a techy-nerdy way of telling the search engines that one particular page is the originator of the information (or the original page where the information came from). Now it doesn't matter if your information has been copied by someone else (at least, to the search engines) because you can tag your page as the canonical page.

This is also useful in ecommerce sites. It is possible, with the dynamic way ecommerce pages come into being, to have the same information about a product you are selling on several different pages.

Google and Bing may decide to list a page that is not the best one (the one with all the great SEO stuff you put on it), so you can tell Google and Bing which is the original page.

Now, putting the canonical link onto a web page requires coding. But, if you are using WordPress, you can use an app like *Canonical Link* to take care of the coding part for you.

Even though you won't get dinged for having multiple copies of the same page on your web site (or other sites) anymore, it's still vitally important to tell the search engines which is the page you want them to use.

SEO Technique: Web Site Speed

"Warp speed, Scottie!" has never been more true. Even though some homes can have Gigabit speeds these days, something only quite large companies could afford just a few short years ago, how fast your web site loads is more important now than ever.

The reason is because of... phones. Over 30% of all web searches are now done on mobile phones (although I think these statistics are skewed a bit. Why? Because these statistics think that "call home" is a search term) and mobile phones are about 70% slower than wireless LAN settings (Wireless is about 50-80% slower than hard wired, in case you were interested).

So, Google decided to make "mobile-first" web sites a priority. How fast is you web site? I'm so glad you asked!

Go to ww.portbell.com and, under cheat sheets, download the workbook "Is Your Site SEO Friendly?". That workbook will send you to several internet sites so you can learn about your web site's speed, stability, and much more.

Your Timely SEO Strategy

I've told people for years, SEO is not brain surgery, but it is rocket science. A brain surgeon may have an idea of how to fix a problem, but they won't really know until they get inside and look around. But, with rocket science, its; do step 1, then step 2, then step 3. If you do the right steps the right way, you will get the right results.

Here are the steps I recommend:

1. Check your existing web site to make sure that the search engines can index it properly. Go to **www.portbell.com** and download the cheat sheet "Is My Web Site SEO Friendly" to find out.

2. Update every page. Add good, high quality, relevant text. Make sure the title and description META tags are there and update them.

3. Review your blog posts for the last year. Are there enough small blog posts (you have been doing weekly small blog posts, right?) that you can combine into a larger post? This is a lot easier than having to create a large new post from scratch (so start doing those weekly posts if you are not already).

4. Determine a good social media strategy. Will it be video, audio, or written (or all 3)?
Here is a good 2021 Secret: Do video posts. Record them in your office then upload them to Facebook, LinkedIn, your blog, YouTube, etc. But you can also strip the audio out and make it a podcast. Then you can take the transcript and make a blog post out of it. So, you can do video, audio and written all with one post.

And let's keep it to those four things. As busy as our schedules are, those three could take months to do.

SEO Secret/Myth: SSL Secure Web Pages

This is an SEO secret that should be a myth. Google, in its infinite wisdom (I'm not sure whether or not I'm being sarcastic) has decided that secure web sites will rank higher than non-secure web sites in their search engine.

———

What is SSL? This gets a bit techy-nerdy so you can skip this part if you want.

SSL, or Secure Sockets Layer, adds a layer of encryption to a web site. That encryption allows you to input text into a form (like a form that asks for your name and phone number) but a hacker will not be able to see that information if he/she is watching internet traffic. They will only see gibberish. The information gets encrypted on one end and decrypted on the other end.

SSL can be used on a web site or in email. Beware, however, if you use it in email. The person you are sending the email to must also have SSL enabled in their email or it won't be secure on their side.

———

The security I'm talking about is that little lock you see at the top of the page, near the URL of a web site, like this:

Portbell.com is secure.

Intra-designs.com is not secure

Here is the rub. Neither of these web sites collects any secure information, so *why should it be important to have that security?*

It isn't. But, Google has decided that they want it, so we bow to them in order to get our web sites higher in the search engines.

Go to **shop.koolwebhosting.com** if you would like to get an SSL secure certificate for your web site. Look for SSL under Security.

SEO Myths

I hope the SEO Secrets got your heart pumping and gave you renewed hope about getting to the top of the search engines.

I have come across many people saying many things over the course of my time as a SEO entrepreneur, and some of the things I've hear are pure myth. Here are a few of them:

SEO Myth: My Customer is Anyone

What do customers have to do with SEO secrets and myths? Well, the secret is that you can't do search engine optimization without knowing who your customer is. Well, you can, but it would be like putting out a fire with $100 bills; it can be done but it's not worth it.

This myth deals with keywords. For those of you who think: "My customer is anyone," that's a myth. Like I said about the plumber earlier; if you're in Florida, you don't care if someone in Montana finds you because you're not going to drive to Montana to fix a clogged sink. And I doubt that they would want to pay the extra, what, ten-twenty thousand dollars in time and gas for your big trucks.

Your customer is not anyone or everyone. Really pinpoint who your target market is. This is good for marketing your company and for search engine optimization. It's flat-out good for you because when you use specific keywords you will get clients that want to purchase what you are selling. Just like the restaurant owner whose target market was Pacific Avenue; he specifically wanted Pacific Avenue keywords. He used other keywords as well, but he knew who his customer was and so he targeted his specific locality and his keywords for that customer.

SEO Myth: Social Media

(Also read the Social Media section under SEO Secrets)
"Social media helps my website." Myth. Social media is changing, almost on a daily basis right now, isn't it? It's changing almost as fast as computers used to change and you can make money off advertising on social media. But social media does not help with SEO on your website. Don't get me wrong; it might help you make money; it might help your business, but it won't help you with search engine optimization.

People might find your Facebook page on the search engines; but they're going to send people to your Facebook page, not your website. Facebook pages are just people talking about stuff and sometimes you might have interesting stuff on your Facebook page, but it is Facebook, not your web site.

And hey, if your Facebook page is your website, put your big-boy pants or big girl pants on and get a real web site; it's worth it. I know everybody thinks that Facebook is here to stay but look at what's going on.

Twitter, Instagram, Pinterest, Snapchat, and all the other "new best things" are all trying to outdo Facebook. Facebook is not going to be at the top forever.

And if you think it will be here forever, do you remember MySpace? How about Prodigy or America Online? They used to be huge, but not anymore.

Even Google+ went down in 2019, and Google has more money than Facebook. I know businesses that concentrated all their marketing on Google+. Guess where they are now. Correct, they are in the trash right next to Google's failed social media attempt.

Get your own website. All the way back before Prodigy they had web sites. Web sites are still what people look for when they do a search; and search engines look for websites before social media sites.

If I'm looking for something and I see a link to a Facebook page, I don't generally click on it, do you? Most people don't. So, follow the logic:

You only have a Facebook page for your small business. Most people will not click on your businesses Facebook page, so you just lost out on a lot of possible customers.

Have a quality web site designed with a WordPress blog (see the first SEO Secret: Blogs). Go to **www.koolwebhosting.com** if you don't have a blog. Grab your domain name, hosting and blog all in one place.

Then, when you post to social media, link to your web site or your blog. That way, when read your posts and want to know more, they have somewhere to go.

SEO Myth: Small Sites Cannot Compete

There are some very big hitters that spend a lot of money on search engine optimization and that may make smaller companies think that they cannot compete in the SEO arena, especially in Google.

But that is not true. A well-designed web site that fits a niche can outperform a bigger site. As an example, a small up-and-coming attorney stared his practice a few years ago, but he was against some heavy competition in the search engines. Attorney keywords are some of the most competitive in the world.

So, I worked with him and helped him target his idea client down to the head of a pin. Then we redesigned the site to fit that market and he was number one in Google within 8 months. I know that sounds like a long time, but it isn't considering all the competition.

Small companies can compete when working with niche keywords. The trick is to find the right keywords. I worked with one company that got to page one on Google within 6 weeks. This company had a lot of national competition, but very little local competition.

Consider this: Web designers can work for anyone anywhere, so they generally try to get to the top of the search engines for keywords like "web design," "web development" and such. But they are competing with every other web designer in the world.

Why not start locally? A Seattle, WA web design company can get to the top of the search engines with a keyword like "Seattle web design" a lot faster than national keywords. And rather than competing with a thousand competitors, they are competing with less than 50.

SEO Myth: The H1 Conundrum

The word is out, you can now use more than one H1 tag on a web page, but DON'T. If you are not familiar with H1, here you go: Back in the old days when they had tree-killing things called newspapers, they had different types of headings. H1 was the most important heading on a newspaper page, and there was only one H1 per page. There could be several H2's, but only one H1.

This was adopted by search engines when web sites got popular. However, since more and more people are making web pages without thinking of the proper coding, the search engines have changed. **But this will hurt your web ranking**. Having more than one H1 tag on your web site means the search engines will not know what's most important for that page, so they won't give a high priority to ANY of the H1 tags. It is still vitally important to have only one H1 tag per page.

SEO Myth: Google Page Rank

"Google Page rank is a must."

Google rates pretty much every page that is listed in its search engine. Those scores are not available to the public as of 2016, but the damage has been done. Too many people think that Google's page rank is a big deal.

I've heard this over and over and over again. I've been hearing this since about 2002, because Google wasn't around in 1995.

There were other websites around back then. Altavista, Lycos and Infoseek were big back then; and there was the little upstart, what name was it? Oh yeah, Yahoo was just starting up. They hadn't even had their initial public offering yet.

Google started getting big around 2000 when they came up with their Page Rank system. And, ever since then, I have been hearing that Google Page Rank is a must. Or, to be more specific, your web site is like a fly on cow dung if your Page Rank is below a 5 (out of 10).

This is a myth. It was brought on by, I hate to say it, SEO "experts" that were out to get money from unsuspecting people.

Basically, you get a Page Rank of zero the day that Google finds your web site on the Internet, and the day after that you get a Page Rank of one. 10 is the highest available page rank, and the Google search engine itself is one of the few web sites that enjoy that top position.

I have had web sites with the Page Rank of one that outperform web sites with a PageRank of six. My best performing web site ever got to a page rank of three. And yet, I have put over 3,000 keywords onto page one of Google for over 200 clients. So, Google Page Rank is not a must.

It does help, don't get me wrong. Google's Page Rank is determined by many factors (and Google has never told everything that the page rank system looks at). Some of the things Page Rank looks at are:

- How many people are going to the website.
- How many other websites have links to that web site (backlinks).
- Quality of content.
- How well the page ranks on the W3C (an independent consortium that decides on HTML "standards").

- And more.

Page Rank is no big deal, however. You will automatically gain page rank as you do SEO work on your site and, like I said, getting a high page rank is not important.

SEO Myth: YouTube

"YouTube videos help with SEO." YouTube videos do absolutely nothing to help with SEO, which may be surprising to you if you already know that YouTube is the second largest search engine in the world right now (which is probably why Google owns it). But YouTube videos will not help with search engine optimization in the least bit.

This is another myth that may change. The reason why it's a myth right now is because there is no speech-to-text recognition on a YouTube video (for techy-nerds, don't get into the cc. You know it's not good at all).

Reliable sources told me that Google has been working on quality speech recognition for some time for YouTube videos, but there are so many accents that it has been difficult.

So, as soon as they perfect it, this will no longer be a myth. But, as of 2020, it still is a myth.

For now, when you do a video, the search engines don't care in the least bit about it; which is why the title, the keywords, and the description are just as important for YouTube as it is for your web site, as it is for your WordPress blog. If you skipped straight to this section, go back and read the SEO Secret on Meta tags.

YouTube is the second largest search engine because of those tags. Google has little idea what's in the video, so putting accurate descriptive tags tells Google what that video is about. You can also use hash tags in the description.

SEO Myth: Mobile-Friendly

Myth: Web sites must be mobile-friendly, or they won't be indexed by the search engines.

This is another myth that may not be for much longer. However, as of the publishing of this book, it is still a myth.

Most web sites made since 2015 are responsive web sites; meaning they automatically change depending on whether you are using a smart phone, tablet, notebook or desktop.

And fear-mongers will tell you that your web site will not rank in Google if it is not responsive. It is a way for them to get more money from you.

The truth is that a web site *not* designed to change for smaller devices will not rank *as high* in the *mobile* search engines as sites that are responsive. Your old, 2005 web site will still rank the same on the desktop version of search engines, for now.

UPDATE: This may change soon. Google is now doing "Mobile-First indexing," making mobile-friendly web sites more important (SEE the new SEO secret: "Mobile-Friendly"). They are, for now, still distinguishing between searches done on a desktop computer and mobile device, but who knows for how long? Get your web site mobile-friendly.

Techy Terms

Some of the terms in this book are kind of techy, so here is my definition of some terms for you:

Backlinks: The internet is built on links. Every time you click a highlighted portion of a web page, or an image, you are clicking on a link.

A backlink is a link from some other web site that links to you (links back to you, hence a backlink).

See the SEO Secret: Backlinks for more information.

Blog: Most people think that blogs just came out around 2005-2010, but blogs are actually very old; one of the oldest ways of communication on the internet.

The modern definition of Blog is a "web log," but that's not the original definition. Way back, before the modern internet; before Prodigy and America Online (yes, that far back, almost to dinosaurs), blogs were used by college professors and students to communicate about projects and dissertations. The original definition of a blog was a "bulletin-board log," hence, Blog.

See the SEO Secret: Blogs for more information.

Domain Masking: allows more than one domain name to point to the same web site without letting people know that they are both pointing to the same web site.

Cross Linking: Read the "backlinks definition first. A crosslink is a where your web site links to another web site, and their site links to yours also. This is the best kind of linking.

See the SEO Secret: Crosslinks for more information.

Domain Name: A domain name is the URL of your web site. For instance, our main site is **www.portbell.com**. The domain name would be portbell.com.

And the domain name can be completely different from your company name.

We have a marketing web site with the doman name tacomaseo.com. You can make a domain name any name, or any series of names that you want (as long as someone else doesn't already have it).

Go to **www.koolwebhosting.com** to explore different types of domain names and see if they are available.

See the SEO Secret: Hello! My {Domain} Name Is... for more information.

Freelancer: A person who works on their own doing a specific task. You could hire a web design company or hire a freelancer to design a web site for you. Freelancers are usually found in 3rd world countries (India, Pakistan, etc.), but there are freelancers everywhere.

Another word for a freelancer could be a solopreneur.

See the SEO Secret: Freelancers for more information.

Hosting: A web site needs an address just like a house does. Web hosting is the place on a computer server where your web site files resides. And the internet knows your web site verses other people's web site by its address. For the internet, it is an IP (Internet Protocol) address, a series of up to 12 numbers with a series of three periods in between. For instance, an IP address might be 192.168.0.1.

See the SEO Secret: Hosting for more information.

IP Address: See Hosting above.

Keywords: Keywords are the words you want to be at the top of the search engines. "Keywords" is a bit misleading; the more proper word would be "keyphrase," because a keyword can be more than one word. For instance, if you are a plumber in Tonga, a great keyword would be "Tonga plumber."

See the SEO Secret: Keywords for more information.

Marketing Web Site: A web site that is only 1-5 pages long and is designed to market to a specific audience. That audience may be for a specific product or a specific location.

Meta tags: These are words, phrases, and other content on the back end of a web page that tell the internet information. There are many types of Meta tags, and I won't go into all of them here (mostly because they change all the time). Do a search on Meta tags for full information.

See the SEO Secret: Meta Tags are (not) dead for more information.

Mobile (phone): Miniputers (let's face it, they stopped being mobile phones years ago) are computers with tiny screens that you can also use to talk to someone.

Page Name: Every web site has at least one page. If there isn't at least one page, it's not a web site. A blog is considered to be a page (if nothing else is available) and even a blog can be many pages (each blog post is considered to be its own page). To easily see a page name, go to that page in a browser and look at the address at the top. You should see the domain name (see above), followed by whatever directories there are, then the page name at the end.

If the page name has weird characters on it, it has been dynamically created. This is common in shopping carts and ecommerce stores.

Social Media: This is a widely encompassing term. It is a reference to any and all social media sites.

SSL: (Secure Sockets Layer) is the lock you see in the address bar of a web site letting people know it's secure to put their private information on that site.

See the SEO Myth: SSL for more information.

Website (or Web Site): A place on the internet where you have information that you want to give other people. It may be information about your company, your funny cat, an organization, stuff you want to sell, or a place to record your bowel-movements. It is completely up to you what you want to put on your web site and there are very few things that will get you in trouble with the law.

Search engines, on the other hand, will rank your web site on its authenticity, legality, and morality. You will notice that pornographic web sites do not commonly come up in searches.

It is good to keep your web site rated PG if you intend on selling normal products and services.

Whitepaper: an official document by a company that is viewable to the general public from their web site.

Contact Me

Yes, the secrets and myths contained in this book have very little coding, but some of it is techy-nerdy stuff. I know it can be overwhelming for some people. If you need help, use the contact form at www.portbell.com. Let me know you read this book. I would be happy to talk to you about whether we might be a good fit for doing some SEO consulting work for you.

Always remember (and this has nothing to do with SEO secrets): Business is never good or bad out there; business is good or bad in your own head first.

I've seen some of the best business ideas fail because the owner simply didn't believe that they can do it.

You can do it; I know you can do it and I would love to be able to help you see that you can do it.

If you're stuck; you don't know where to go, you're banging your head against the wall, can't figure something out and you don't know why you can't figure it out… whether it's your business, career, or your life and you can't figure out why you're hitting that glass ceiling, contact me. Go to www.portbell.com and fill out the contact form.

And while you are there, be sure to join my Success Tips newsletter and keep updated on blog posts, books, and other updates.

Thank you for reading this book, I hope it gave you some great information.

Special bonus! Watch the original webinar by going to:
www.webishops.com/supersecret.html

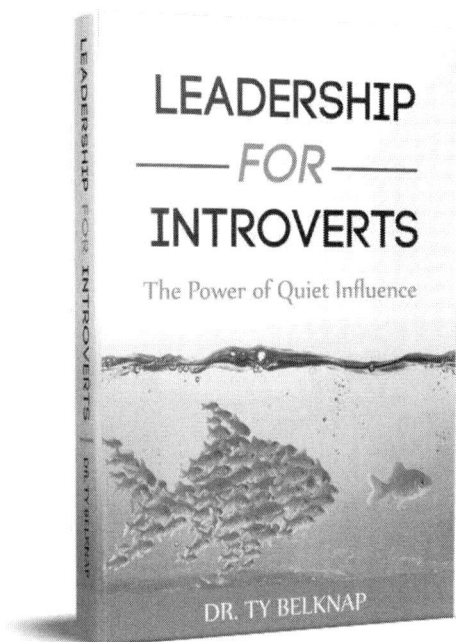

Leadership for Introverts Now Available
www.leadershipintroverts.com

Printed in Great Britain
by Amazon

85046990R00063